Concept Sciences, Incorporated
Hanover Township, Pennsylvania

Investigated by: John Lee Cook, Jr.

This is Report 127 of the Major Fires Investigation Project conducted by Varley-Campbell and Associates, Inc./TriData Corporation under contract EMW-94-4423 to the United States Fire Administration, Federal Emergency Management Agency.

Department of Homeland Security
United States Fire Administration
National Fire Data Center

U.S. Fire Administration Fire Investigations Program

The U.S. Fire Administration develops reports on selected major fires throughout the country. The fires usually involve multiple deaths or a large loss of property. But the primary criterion for deciding to do a report is whether it will result in significant "lessons learned." In some cases these lessons bring to light new knowledge about fire--the effect of building construction or contents, human behavior in fire, etc. In other cases, the lessons are not new but are serious enough to highlight once again, with yet another fire tragedy report. In some cases, special reports are developed to discuss events, drills, or new technologies which are of interest to the fire service.

The reports are sent to fire magazines and are distributed at National and Regional fire meetings. The International Association of Fire Chiefs assists the USFA in disseminating the findings throughout the fire service. On a continuing basis the reports are available on request from the USFA; announcements of their availability are published widely in fire journals and newsletters.

This body of work provides detailed information on the nature of the fire problem for policymakers who must decide on allocations of resources between fire and other pressing problems, and within the fire service to improve codes and code enforcement, training, public fire education, building technology, and other related areas.

The Fire Administration, which has no regulatory authority, sends an experienced fire investigator into a community after a major incident only after having conferred with the local fire authorities to insure that the assistance and presence of the USFA would be supportive and would in no way interfere with any review of the incident they are themselves conducting. The intent is not to arrive during the event or even immediately after, but rather after the dust settles, so that a complete and objective review of all the important aspects of the incident can be made. Local authorities review the USFA's report while it is in draft. The USFA investigator or team is available to local authorities should they wish to request technical assistance for their own investigation.

This report and its recommendations were developed by USFA staff and by Varley-Campbell & Associates, Inc. Miami and Chicago, its staff and consultants, who are under contract to assist the Fire Administration in carrying out the Fire Reports Program.

The U.S. Fire Administration greatly appreciates the cooperation received from Director John Conklin of the Lehigh County Emergency Management Agency, Stephen Martin of the Eastern Pennsylvania EMS Council and David Lesak. Appreciation also goes to Scott Snyder, Pennsylvania State Police and Chief Robin Yoder of the Hanover Lehigh County Fire and Rescue Company.

For additional copies of this report write to the U.S. Fire Administration, 16825 South Seton Avenue, Emmitsburg, Maryland 21727. The report is available on the Administration's Web site at http://www.usfa.dhs.gov/

U.S. Fire Administration

Mission Statement

As an entity of the Department of Homeland Security, the mission of the USFA is to reduce life and economic losses due to fire and related emergencies, through leadership, advocacy, coordination, and support. We serve the Nation independently, in coordination with other Federal agencies, and in partnership with fire protection and emergency service communities. With a commitment to excellence, we provide public education, training, technology, and data initiatives.

TABLE OF CONTENTS

Concept Sciences Incorporated
Hanover Township, Pennsylvania
February 19, 1999

Investigated By: John Lee Cook, Jr.

Local Contacts:

John Conklin, Director
Lehigh County Emergency
    Management Agency
455 West Hamilton Street
Allentown, PA 18101
610-782-3073

Scott M. Grim, Coroner
County of Lehigh
501 West Hamilton Street
Allentown, PA 18101-1614
610-782-3426

David M. Lesak
Hazard Management Associates
PO Box 3004
Allentown, PA 18106
610-395-6409

Stephen J. Martin, Assistant Executive Director
Eastern Pennsylvania EMS Council
1405 Cedar Crest Blvd., Suite 208
Allentown, PA 18104
610-820-9212

Scott R. Snyder, Lieutenant
Pennsylvania State Police
2930 Airport RD
Allentown, PA 18017-2149
610-861-2100

Robin Yoder, Chief
Hanover Lehigh County Volunteer Fire &
    Rescue Company
1001 Postal RD
Allentown, PA 18103
610-264-7227

OVERVIEW

On Friday, February 19, 1999, a devastating explosion destroyed a plant operated by Concept Sciences, Incorporated (CSI) in Hanover Township, Pennsylvania. Hanover Township is located in Lehigh County near Allentown. The blast killed five people, caused approximately $5 million in damages, and disrupted electrical service to approximately 1,188 PP&L customers. All of the victims were men between the ages of twenty-five and fifty-five. Four of the deceased were CSI employees and died from the effects of the massive explosion. The fifth victim, an employee of a business located next to the CSI facility, died as a result of a severe head injury from flying debris. One of the deceased CSI employees was a member of an area volunteer fire company.

Fourteen people, including five firefighters, were transported to area hospitals during the incident. The most severely injured were two CSI employees who were buried beneath debris. They were rescued by firefighters more than one hour into the incident. Firefighter injuries included chemical burns, chest pains, lacerations, strained muscles, and shortness of breath.

The explosion registered 0.7 on the Richter Scale at Lehigh University's seismograph center, which is located approximately five miles from the site in Bethlehem. The University's readings indicate that the explosion occurred at 20:14:43 hours and caused the ground to move both up and down and side to side. The explosion could be seen for seven miles and could be felt as far away as Lehighton and Tobyhana to the north and Trexlertown and Longswamp Township to the west. There were numerous reports of windows being blown out of homes on Dauphin Road and Irving Street in Hanover, Township.

The explosion produced a white cloud that rained chemical residue onto the streets of Allentown, approximately two miles from the blast site. An evaluation of the chemical residue concluded that the airborne materials were not harmful and could easily be neutralized with soap and water.

The force of the explosion produced a crater four feet deep and approximately eighteen feet in diameter in the floor of the CSI plant, destroyed three twenty-five-foot-high exterior concrete walls, and partially destroyed the building's concrete roof. Eleven buildings adjacent to the CSI Plant also sustained damage and a number of automobiles were damaged when the exterior wall of a building, which faced the CSI facility, collapsed onto the automobiles and crushed roofs, hoods and windshields.

The cause of the explosion is still under investigation. According to local authorities, CSI employees were in the process of distilling a diluted form of hydroxylamine at the time that the explosion occurred. Authorities indicated that this was the first production run of this material at the plant. The product was used to clean semiconductors and to manufacture pharmaceuticals and CSI was the first US manufacturer to produce the product.

During the peak of the incident over 400 emergency personnel were involved in the incident and included fire and EMS agencies from three counties. A number of local, State, and Federal agencies also assisted the fire and rescue personnel in their efforts. A list of the agencies that responded to the incident is included as Appendix A. Emergency personnel remained at the scene until Monday, February 22, 1999.

KEY ISSUES

Issues	Comments
Early identification of the hazardous material(s) involved in the incident	The first emergency personnel to arrive at the incident did not know what material(s) were involved in the incident. It was 60 to 80 minutes into the incident before the hazardous materials team became fully operational and was able to conclusively identify the product. The explosion destroyed the MSDS's and floor plan of the facility and there were a large number of unmarked 55-gallon drums scattered throughout the blast area.
Personal protective clothing and equipment	The first emergency personnel took pH and oxygen readings which indicated that the area was safe. Operations personnel did not know what material(s) were involved and were unsure of the proper level of protection required. A number of EMS personnel and search dogs entered the area without proper protection. Fortunately there were no serious results of these actions. Some responders also purposely avoided being decontaminated in order to remain active in the incident. Personnel did not have a secondary set of personal protective clothing.
Communications	The incident involved agencies from three counties. There was no set of frequencies at the operations level that were common to all the agencies involved in the incident. The incident also generated approximately 400 calls to the 9-1-1 Center, which was staffed by six personnel.
Coordination	More than 400 personnel responded to this incident. Agencies from local, State, and Federal governments also responded. Unified incident command and a solid emergency management plan allowed this incident to be managed to a satisfactory conclusion.
Clean-up operations	CSI hired a private contractor to cleanup the site of the explosion. Work was delayed more than twelve days because insurance companies would not approve the work. The State assumed the responsibility and began work on the fourteenth day following the incident. Cleanup involved run-off control, removal of structural and other debris, and handling of the drums containing the hydroxylamine solutions.
Time of day	The incident occurred on a Friday evening. This minimized the number of injuries and fatalities because many of the buildings were unoccupied at the time of the explosion. Two nearby day-car centers were also closed. Most of the responders were volunteers and the fact that the event occurred at night and during a weekend no doubt enhanced the number of people who were able to respond and participate throughout the duration of the event.

HANOVER TOWNSHIP

Hanover Township is located near Allentown in Lehigh County, Pennsylvania. Lehigh County's population is approximately 350,000. The Township is home to approximately 2,500 people and encompasses five square miles. The Lehigh Valley International Airport covers nearly one half of the Township's total land area. Fire service is provided by the Hanover Lehigh County Volunteer Fire and Rescue Company Number One. The department was founded in 1978 and has forty members, all of whom are volunteers. The Department responds to approximately 300 incidents per year.

The fire department has one station, designated as Station 33, and maintains an engine company (E 3312), a 75 ft. quint (Ladder 3331), a rescue-pumper (E 341), a tanker (Tanker 3321), and a rescue squad (Rescue 3341). The Northampton Regional Emergency Medical Service provides ambulance service in the Township.

The Lehigh County Hazardous Materials Team provides hazardous materials response to the Township. The team, established in 1990, is an agency of the Lehigh County Government and is composed of seventy paid-on-call members, many of whom are local career and volunteer fire and rescue personnel.

BUILDING HISTORY AND OCCUPANCY

The Concept Sciences, Incorporated facility was located at 749 Roble Road in the 1,600-acre Lehigh Valley Industrial Park. A map of the complex is attached in Appendix B. The industrial park is southwest of the Lehigh Realty Associates of Roseland, New Jersey and had been occupied by CSI in September 1998.

CSI began their operations in April 1997 in the Allentown Business Park and was reported to be the first, and only, manufacturer and supplier of free-base hydroxylamine in the United States. The product was used by pharmaceutical companies and to clean semiconductors. Known as 50-HA, the hydroxylamine is a clear, colorless liquid in the form of a fifty percent solution in water. The NFPA rates the hazards of hydroxylamine as: Health-2, Flammability-0, and Reactivity-3. A Hazardous Materials Guide for the product is attached as Appendix C. The MSDS published by CSI, which is included as Appendix D, listed the product's hazards as:

- Corrosive

- Corrosive to skin – causes burns

- Do not distill to dryness

- Excessive and extended heating may cause a "pressure build-up" explosion within a closed container

- Decomposes in presence of heat and air. Decomposition is accelerated with transition metal ions (e.g. Fe, Cu, Ni, et al)

There are approximately 2,000 gallons of the finished product on site and thirty to forty blue, unmarked 55-gallon drums. Very little of the materials was found after the explosion and it is theorized that most of the material was consumed during the explosion. The sprinkler system ruptured and water flowed from the broken pipes for a period of time causing a run off problem in a nearby creek. The Incident Commander indicated that some of the product apparently made its way into the sanitary sewer system and the local sewer plant recorded some unusual readings as a result.

BUILDING CONSTRUCTION

CSI occupied 19,200 square feet on the north end of a 42,000 square foot building. The site plan is attached as Appendix E. The one story building had a ceiling height of approximately twenty feet. Constructed in 1987, the noncombustible building had a steel frame and the building's exterior walls were constructed of concrete masonry units (CMU's). An interior CMU firewall separated the CSI facility from the other occupancies within the building. The firewall was penetrated by debris from the explosion and resulted in the death of an occupant in the leased space immediately adjacent to CSI.

The building was fully sprinklered and the system was operational at the time of the explosion. The explosion destroyed the system in the CSI plant and water was pouring into the facility when the fire department arrived at the scene. The system was shut off by the fire department shortly after their arrival when it became evident that runoff was occurring from the site. The occupancy did not have a fire alarm system according to the New Tenant Inspection Report obtained from the Hanover Township. The Township had adopted the BOCA Building Code.

A key to the building was provided for fire department use, as required, and was enclosed in a rapid entry lock box located near the main entrance. Emergency information, including MSDS sheets about the chemicals used in manufacturing process and floor plans, was provided inside the building near the main entrance. The explosion destroyed these materials and this information was not immediately available to the fire department. Therefore, the fire department did not immediately know what materials were involved in the explosion and resulting fire.

THE EXPLOSION

The explosion registered 0.7 on the Richter Scale at Lehigh University's seismograph center, which is located five miles from the CSI Plant in Bethlehem, Pennsylvania. The University's readings indicate that the explosion occurred at 20:14:43 hours on Friday, February 19, 1999, and that the blast caused the ground to move both up and down and side to side. The explosion was seen as far as seven miles away and was felt as far away as Lehighton and Tobyhana to the north and Trexlertown and Longswamp Township to the west.

There were numerous reports of windows being blown out of homes on Dauphin Road and Irving Street in Hanover, Township. The blast produced a white cloud that rained chemical residue onto the streets of Allentown, two miles away. A subsequent evaluation of the chemical residue concluded that the airborne material was not harmful. The white cloud and the magnitude of the explosion generated a significant amount of interest in the residents of the area.

The Lehigh County 9-1-1 Center received approximately 300 to 400 calls during the first few hours of the incident. There were six personnel on duty at the time of the explosion and the unusually high volume of calls overloaded the Center. Reports of an explosion were also received by other agencies in the vicinity, including the Lehigh Valley International Airport.

The Fire Chief of the Hanover Lehigh County Volunteer Fire and Rescue Company Number One, the first due fire company for the CSI facility, was in his vehicle and was monitoring his scanner when the explosion occurred. The chief immediately responded in the direction of the industrial park based upon the information that he was hearing. While he was enroute, an alarm was transmitted at 20:22 hours for a reported explosion at the TruGreen Chemlawn Facility, located in the industrial park at 764 Roble Road.

At 20:25 hours, the Communications Center confirmed that the correct address was 749 Roble Road based upon reports from a security service. At 20:29 hours, the first two EMS units were dispatched.

The fire chief was the first fire official to arrive at the scene. Several members of the Pennsylvania State Police met the chief at the scene; having responded from their barracks located a few blocks from the site. They were asked to evacuate the buildings across the street. The fire chief gave an initial size up and reported that approximately one-half of the building had collapsed and that there was damage to several adjacent buildings. A search of the area was ordered at 20:33 hours and command was established at 20:35 hours. The Command Post was initially the chief's car, but was transferred to the county's mobile command post later in the operations.

The incident commander promptly began to summon additional assistance. There is no standardized multiple alarm system in effect in Lehigh County. Therefore, the IC requested specific agencies to respond based upon his personal knowledge of their capabilities and resources. The initial request asked for companies from Allen Township, East Allen, Hanover, Catasauqua, and Northampton Regional EMS.

Throughout the duration of the incident, the fire chief managed all of the requests for fire and EMS resources. The county's Emergency Management Director responded to the Command Post early in the incident and assisted the incident commander by coordinating the requests for all other resources.

In accordance with the Department's SOPs, the firefighters from the Hanover Lehigh County Volunteer Fire and Rescue Company Number One (Station 33) respond to their station to staff their apparatus rather than responding directly to an incident. Given the close proximity of the incident to their station, this practice did not delay their response and actually helped reduce congestion at the scene that is often created by the presence of the volunteer emergency responders' personal vehicles.

Apparatus from Station 33 began to arrive at 20:32 hours. Firefighters arrived on the scene to find a collapsed building, smoke drifting from the blast sight, and a small fire of little real significance. Water was pouring from the demolished automatic sprinkler system and there was debris everywhere. Ceiling lights were hanging down in a number of buildings and a large number of spectators had assembled due to the magnitude of the blast. Law enforcement agencies quickly secured the scene and the utilities were subsequently shut off.

The weather at the time of the explosion was very cold. The temperature was in the low twenty's (Fahrenheit) and there was a wind from the northwest that created a wind chill in the single digits. Whenever the temperature is below freezing and there are windy, wet conditions emergency personnel must take adequate precautions to prevent hypothermia. An excellent reference to consult is Emergency Incident Rehabilitation published by the United States Fire Administration.

The explosion created a crater four feet deep and approximately eighteen feet in diameter in the floor of the CSI plant, blew out three of the building's 25-foot high concrete walls, and partially destroyed the concrete roof. The explosion damaged eleven other buildings, including the RPS package delivery service located at 759 Roble Road, Sugarloaf Mid-Atlantic, and the Meixell Brothers Warehouse.

The exterior walls of RPS and Meixell Brothers that faced the CSI plant crashed onto the automobiles in the parking lot, crushing roofs, hoods and windshields. A glass door was also blown through at Nikon Precision, Inc. located across Roble Street from the CSI facility. Much of the damage to the adjacent structures appears to have been caused by the negative pressures created by the explosion.

The Northampton Regional Emergency Medical Service provides EMS transport service to Hanover Township. The first unit from the agency established command of the medical operations at 20:39 hours. EMS Command initially established five different triage areas due to the size and scope of the incident. Because there were multiple casualties at the incident, EMS operations were conducted in accordance with the Disaster Operating Guidelines promulgated by the Eastern Pennsylvania Emergency Medical Services Council. Radio communications for the EMS operations were coordinated by MEDCOM, a service that connects the sixteen area hospitals with the local area rescue squads.

At 20:40 hours, rescuers confirmed that there was an entrapment at the site. At 20:46 hours, a level one disaster was declared indicating that at least six people were trapped in the debris. Rescue efforts were conducted on both the East and West sides of the building and sectors were established in each area. There was less damage on the East Side and the North wind kept this sector relative free of smoke. Conditions were more difficult on the West Side, however. The adjacent buildings that had been damaged by the blast were also searched several times.

Fourteen people, including five firefighters, were transported to area hospitals during the incident. The most severe injuries were sustained by two CSI employees found buried beneath debris on the East Side of the collapse zone by firefighters more than one hour into the incident. Firefighter injuries included chemical burns, chest pains, lacerations, strained muscles, and shortness of breath. The EMS commander initially thought that the incident was a natural gas explosion and did not order the decontamination of the first victim to be transported.

The explosion killed five men, aged 25 to 55. Four of the deceased were employees of CSI, including two victims (father and son) from the same family. Ironically, the father was a long time member of a local volunteer fire company. According to the coroner's report, the CSI employees were all killed as a direct result of the effects of the blast. The fifth victim, an employee of Sugarloaf Mid-Atlantic, died as the result of a severe head injury.

The first emergency personnel to arrive at the incident did not know what material(s) were involved in the incident. It was initially believed that the blast had resulted from an explosion of natural gas. The force of the explosion destroyed the MSDS's and floor plan of the facility and there were a large number of unmarked 55-gallon drums scattered throughout the blast area. Oxygen and pH levels were taken and it was felt that personnel could safely operate in the area if they used bunker gear and SCBAs. The situation was further complicated by the fact that a representative from CSI downplayed the dangers of the material involved saying that it was merely corrosive and could easily be neutralized with soap and water.

A number of the mutual aid companies were concerned about the presence of hazardous materials and staged some distance from the building and requested orders from Command before they deployed their personnel. The sector commanders also began to become concerned during their search and rescue efforts because they were finding chemical residue, which indicated that hazardous materials were present. They contacted Command about their concerns and it was felt that bunker gear and a SCBA would provide sufficient protection to rescue personnel.

The Lehigh County Hazardous Materials Team was requested at 20:30 hours and responded to the incident. The team, however, did not become fully operational until approximately sixty to eighty minutes after the first fire company arrived at the scene. The delay resulted in the first responders not being able to conclusively identify the product until well into the incident. The Lehigh Team was assisted in their efforts by the Hazmat team from Berks County. The two teams established an inner perimeter and a decontamination sector. Approximately eighty people and several search and rescue dogs were decontaminated during the incident. Even though the dogs were not allowed in the inner perimeter, several dogs sustained glass cuts and chemical burns to their paws.

The majority of the emergency responders at the scene did not have access to a second set of personal protective clothing. As a result, a number of personnel deliberately avoided going through the decontamination process because the Hazmat team was impounding their protective clothing, which meant they could no longer operate within the inner perimeter of the site.

The Hazmat team decontaminated the five bodies of the deceased. The Chief of Lehigh County Hazmat Team made the decision not to rotate the members of the team that was used to decontaminate the profoundly injured, the deceased, and the body parts that were recovered at the scene. The Chief felt that this would minimize the number of team members that would be exposed to the stress and trauma created by the graphic injuries. The Chief also stated that his exposure to the experiences of others through attendance at the United State Fire Administration's National Fire Academy helped him to manage the incident.

Two medical transport helicopters responded to the scene as well as a helicopter from the State Police. The helicopters were used to conduct an aerial search of the site with an infrared camera in an attempt to locate the victims. No one was transported to a medical facility by a helicopter.

The Pennsylvania Urban Search and Rescue Task Force, based in Harrisburg, was activated and responded to the incident. The team is one of the FEMA USAR Teams and this was the first time that the team had been activated to respond to an incident within the State of Pennsylvania. The USAR Team used search dogs, cameras, and the blueprints of the building to coordinate the rescue and recovery efforts. Several other area search and rescue teams also responded with their search dogs. There was some initial confusion because the local responders did not have any training on how to work with a USAR Team.

A number of agencies assisted with the rehabilitation of the first responders. Given the duration of the incident, there was a concern that if people were allowed to leave they would not return. Command had originally planned to relieve everyone at 06:00 hours. It became by 03:00, however, that everyone was very tired. Relief efforts were pushed up to 04:00 and transit buses were used to transport personnel to Station 33 for Rehab. The low temperature and the fact that the station was less than one mile away from the incident site made the station an ideal site. Personnel were rotated through the station throughout the incident. Key command personnel were rotated on a staggered basis in order to preserve continuity at the scene.

The local Red Cross Chapter provided two food vans at the scene to assist the emergency responders. The Red Cross also opened its facility at 2200 Avenue A in Bethlehem for the family members of anyone involved in the explosion.

Communications proved to be a problem because there were agencies from three different counties at the incident as well as numerous State and Federal agencies. There was no common radio frequency that all of the agencies could communicate on. The Incident Command did have radios for both Lehigh and Northampton Counties and was able to communicate effectively with companies from both jurisdictions. Operational personnel, however, did not have this capability. There was also some incompatibility of hose threads between the various fire departments that responded to the incident. Fortunately, there was very little fire and this did not prove to be a major problem.

Approximately 400 personnel were ultimately used to bring the incident to a conclusion and a lot of private vendors were called in to provide special assistance. The Post Office parking lot near Station 33 was used to stage personnel and apparatus in order to account for everyone and to coordinate the rescue and recovery efforts. The parking lot was also used as a staging area for the large number of media personnel who responded to the event. Since Roble Road is a dead end street every effort was made by the staging officer to prevent the influx of emergency personnel and private vendors from congesting the area.

The time of day proved to be an important factor at this incident. If the event had occurred during the normal business day, the number of deaths and injuries would have no doubt been much higher due to the presence of more workers within the park and because there were two day care centers located near the site of the explosion. Additionally, the majority of the fire personnel who responded were volunteers. The large turnout of personnel may not have occurred had the event happened earlier in the day when the majority of the firefighters would have been at work. The event lasted for nearly four days and although companies were rotated during the event, many of the volunteers may not have been able to be absent from their jobs for such a long period of time.

The incident had a profound impact on the community. The emergency responders knew a number of the victims and one victim was a volunteer firefighter with a local fire company. By the third day of the incident, people began to erect crosses in memory of the victims and to bring flowers and prayers to site. This had an impact on the emergency responders who were still at the incident and had not faced this situation before. Stress debriefing was made available to anyone who wanted the service.

A post-incident critique of the event was conducted at 19:00 hours on March 22, 1999 in the auditorium of Dieruff High School, Washington and Irving Streets, in Allentown. The Lehigh County Emergency Management Agency coordinated the process and the session was well attended. The session was taped and the Emergency Management Office published a transcript of the event.

A summary of the Chronology of events is provided below:

Friday 19 February:

- 19:45 CSI workers contact supervisor at home and inform the supervisor that there was a problem, he heads for plant.
- 20:14 An explosion destroys the Concept Sciences Building, damages other buildings, throwing debris more than 200 yards, and cutting power to 1,188 PP&L Inc. customers.
- 20:22 State police and emergency workers rush to scene.
- 20:25 Location confirmed as 749 Roble Rd.
- 20:29 EMS is dispatched.
- 20:33 Search of the site is ordered.
- 20:35 Command established.
- 20:39 EMS Command established.
- 20:40 Entrapment is confirmed.
- 20:45 Lehigh County Emergency Management Services Director John Conklin issues a warning advising people to remain indoors because of potential exposure to hazardous chemicals.
- 20:46 Level One disaster declared.
- 20:50 First live victim recovered from debris.
- 21:00 The Lehigh Valley Chapter of the American Red Cross opens its headquarters at 2200 Avenue A in Bethlehem for the victims' relatives and friends.
- 21:30 The Lehigh County Hazmat team arrives and begins decontaminating people who were exposed to chemicals.
- 21:50 The Incident Commander calls for a crane to remove debris.
- 22:30 Rescuers decide that it is too dangerous to enter the building.
- 22:44 Electric power is restored to area customers.

Saturday 20 February:

- 00:20   Wearing SCBA, firefighters begin searching the wreckage for victims.

- 01:30   Members of the Northeast Search and Rescue Team and East Penn Search and Rescue Teams, arrive.

- 02:00   Officials hold a news conference and announce that they are still trying to locate three of four victims.

- 03:00   A rescue official visits the Red Cross shelter to update family and friends on their progress.

- 04:00   Operations change from rescue to body recovery.

- 06:00   The body of the last fatality is recovered from the rubble.

- 09:20   A news conference is held and officials announce that there are five dead and thirteen injured as a result of the explosion.

- 09:45   Members of the media are allowed to inspect the site.

- 12:00   Lt. Governor Mark Schweiker arrives.

- 13:00   A news conference is held to discuss the investigation.

THE INVESTIGATION

An investigation was conducted by a number of local, State, and Federal agencies following the incident to determine the cause of the explosion. No official determination had been made at the time that this report was prepared and the investigation is still open.

Within the Commonwealth of Pennsylvania, the State Police are responsible for determining the cause and origin of fires and for investigating all deaths that occur. The CSI explosion occurred within the jurisdiction of Troop M, which maintains a team trained investigators for this purpose. An investigative team was dispatched to the incident and the State Police also assisted with on scene security and crowd control. The Troop Commander responded to the incident as well and assisted with media relations and public information efforts.

The State Police assumed command of the incident upon conclusion of the rescue/recovery efforts and after the incident was stabilized in order to conduct a criminal investigation. The Coroner's Office responded to the incident to assist the State Police with the determination of the cause of death of the five victims.

The U.S. Chemical Safety and Hazard Investigation Board (CSB) visited the site at the request of Pennsylvania Governor Tom Ridge. The new Federal agency is modeled after the National Transportation Safety Board and is a non-regulatory, independent Federal agency that seeks to ensure the safety of workers and the public by preventing or minimizing chemical accidents at industries. The agency agreed to conduct an investigation of the incident, the tenth probe of its kinds since the Board began its work in January 1998. The CSB is being assisted in its investigation by experts from Naval Surface Warfare Center located at Indian Head, Maryland and the agency had not made a final determination at the time that this report was prepared.

On August 11, 1999, the United State Department of Labor issued a news release, which stated that the Federal Occupational Safety and Health Administration (OSHA) had cited CSI for safety violations and had proposed penalties of $641,200. OSHA alleged twenty violations, including eleven willful violations for failure to protect employees from the explosive potential of hazardous chemicals and nine serious violations. Willful violations are those committed with an intentional disregard of, or plain indifference to, the requirements of the Occupational Safety and Health Act and OSHA regulations. A serious violation is defined, as one in which there is a substantial probability that death or serious physical harm could result, and the employer knew or should have known of the hazard.

The infractions included violations of OSHA's process management standard (PSM) and the hazard communications standards, which included allegations that CSI had not provided pertinent information to their employees on the hazards involved in the production process or the explosive nature of the chemical. The willful violations included failure to compile and process safety information; inadequate process hazard analysis and operating procedures; failure to train employees on operating procedures and the physical hazards of chemicals; lack of a pre-startup safety review; process equipment deficiencies; and failure to develop mechanical integrity procedures.

Serious violations included the lack of employee participation in a PSM program, failure to adopt safer work practices, no injury and illness logs for contract employees, inadequate mechanical maintenance training, deficiencies in chemical hazard evaluation procedures, and improper labeling of chemical containers.

OSHA's inspection of the explosion site revealed that the explosion occurred at a 2,500-gallon fiberglass reinforced charge tank containing approximately 750 pounds of hydroxylamine. The tank was being used in the distillation process. Pure hydroxylamine has explosive energy roughly equivalent to that of TNT.

LESSONS LEARNED

1. **Prior preparation increases the opportunity for success.**

   The presence of an effective Emergency Management Agency proved to be beneficial at this incident. Every key player interviewed during the preparation of this report stated that they believed that the local emergency planning process was a key factor in the successful management of the incident. Frequent planning sessions and training exercise, including tabletop disaster drills, helped the emergency responders to know each other and enabled them to know what they could expect from each other. The planning process had also identified the resources that were needed to manage the incident. The management of an incident of this magnitude requires a significant amount of human and material resources and prior planning helps insure that adequate resources will be available on a timely basis. Pre-incident planning is important regardless of the hazards involved, but it is especially important whenever a hazardous material is involved. Unfortunately, in this instance there was very little information available that would have indicated that an explosion, especially of this magnitude, would or could have occurred at this site.

2. **A unified incident management system allowed the operation to run smoothly.**

   The fire chief remained in command throughout the initial phases of the incident. Command was transferred in an orderly manner to the appropriate official as the mission of the incident evolved from a rescue operation, to the recovery phase, and finally into a crime scene. Federal, State, and local agencies coordinated their efforts through a single incident commander, minimizing the potential for conflict and confusion that can accompany an event of this magnitude.

Every successful incident includes items that can be improved upon to help the next incident be managed more effectively. One such item that was identified during this incident is the need to communicate more effectively with the front line personnel. The command post personnel easily transitioned from one mission to another as the incident progressed. Some confusion and misunderstanding, however, was reported at the operations level because personnel believed that they were still operating in one phase (e.g. rescue) when the operation's objectives had actually changed (e.g. victim recovery).

Also identified was the need to provide the Incident Commander with an aide or support person early in the incident. The fire chief reported being overwhelmed with face-to-face communications and he often was unable to hear or to respond to radio messages. An aide could have responded to the radio messages and relayed the incident commander's decisions to the appropriate requesting parties. The failure to properly respond to such requests can cause frustration and create opportunities for free lancing in the absence of clear direction from the incident commander.

3. **When search and rescue dogs are used extreme caution must be exercised if there are hazardous materials involved.**

Extreme caution should be exercised when search and rescue dogs are used in conjunction with an incident involving hazardous materials. It is impossible to provide a dog with the same level of protection as a human. Therefore, dogs should not be permitted in the hot zone or interior perimeter. Chemicals that are caustic to the respiratory tract can severely injure a dog and can permanently reduce or destroy a dog's olfactory abilities. A number of search and rescue dogs had to be decontaminated during the incident due to an exposure to the caustic materials involved in the incident and several dogs were treated for chemical burns to the pads of their feet. The glass that was scattered throughout the site by the explosion also resulted in several dogs being cut. Fortunately, the chemicals involved in this incident were relatively minor and did not do any permanent damage.

4. **A responsible party must be identified.**

A responsible party should be identified whenever an incident involves a hazardous material or posses the potential for a catastrophic event, such as an explosion, to occur. The initial responders did not know what materials were involved and all of the employees at the CSI facility had either been killed or were trapped within the debris. The plant had exploded once. Was there a potential for subsequent explosions? This information may only be available with someone familiar with the operation and who has the authority to make decisions for the parties involved, i.e. the owner or manager.

The manager of the CSI facility was initially reluctant to assume any responsibility due to the obvious liability issues involved and the potential for litigation and possible fines from regulatory agencies. The manager's reluctance significantly delayed the cleanup activities at the site. The location had the potential for another explosion, a fire, a structural collapse, or any combination of these events. The longer it took for cleanup activities to commence the greater the risk that of one or more of these events might occur.

5. An adequate level of personal protection must be provided to emergency responders and they must be properly trained to use the protective devices.

The product involved in the explosion and fire at the CSI plant was not positively identified for over one hour. In spite of this fact, firefighters conducted a search for victims and treated the injured. Firefighters wore their structural turnout clothing and SCBA's. A number of the EMS personnel entered the inner perimeter without any protective clothing or respiratory protection. Fortunately, the hazardous materials involved were rather benign.

The issue of how much protective equipment is required is a significant one. Level C chemical suits and air-purifying respirators would have been sufficient for this incident. The Lehigh County Hazmat team had less than forty Level C Suits. More than 400 people were involved in the incident. Local jurisdictions should re-evaluate their cache of resources and ensure that their stock is sufficient for a major incident. One alternative is to identify local vendors that may be able to supply the suits on a timely basis.

The fire investigators, State police officers, and coroner's personnel who responded to the incident did not have a significant amount of training in handling hazardous materials incidents and they lacked the protective clothing and respiratory protection devices required to safely work at an incident of this type. It is essential that all personnel expected to work in hazardous environments be properly trained and equipped to safely perform the tasks that will be expected of them.

A related matter is the need to collect and decontaminate the personal property of people who are killed or injured in a hazardous materials incident. These items must be properly marked and the chain of custody must be maintained because items may be crucial to the investigation, particularly if the incident is determined to have been criminal in nature.

6. The time of the day can have a significant impact on the outcome of an event.

There are two daycare centers located in the Lehigh Valley Industrial Park and the businesses located within the park have a significant number of employees. If the explosion had occurred during the normal workday it is very likely that the number of deaths and injuries would have been far greater in number.

APPENDICES

APPENDIX A

List of Responding Agencies

The following agencies were involved in the response to the explosion at the Concept Sciences, Incorporated facility. The list was provided by the Hanover Lehigh County Volunteer Fire and Rescue Company and the author of this report regrets any unintentional omission of a responding agency.

Fire Departments:

Alburtis Fire Company
Allen Township, Northampton County
Allentown Fire Department
Berks County Hazmat Team
Catasauqua, Station 2
Cementon, Whitehall Station 40
Cetronia Fire Company
Citizen's Fire Company
Coplay, Station 5
East Allentown Township, Northampton County

Emmaus Fire Department
Fogelsville Fire Company
Fountain Hill, Station 34
Fullerton, Whitehall Station 36
Han-Le-Co, Station 33
Hanover Twsp, Northampton Co.
Lehigh Valley Hazmat Team
Lehigh Valley Airport
Schnecksville Community
Woodlawn Fire Company

Emergency Medical Services:

Allentown, Station 78
Bath, Northampton County
Bethlehem Twsp, Northampton Co.
Cetronia, Station 62
City of Allentown
East Allen Twsp, Northampton Co.
Fountain Hill, Station 73
Hershey Medical Center Lifelion
Lehigh Valley Hospital

MEDEVAC
Muhlenberg Hospital
Northampton Regional, Sta. 75
Northern Valley, Station 67
Sacred Heart Hospital
Salisbury, Station 68
St. Luke's Hospital
Suburban EMS
Upper Saucon, Station 72

Law Enforcement Agencies:

Catasauqua Police Department
City of Allentown Police Department
Fountain Hill Police Department

Lehigh Valley International Airport PD
Pennsylvania State Police
Whitehall Police Department

Appendix A (continued)

Government Agencies:

Department of Environmental protection
Federal Emergency Management Agency
Hanover Township
Lehigh County Assessment Office
Lehigh County Coroner's Office
Lehigh County Emergency Management Agency
Lehigh County Public Affairs Office
Northampton County Emergency Management Agency
Occupational Health and Safety Agency
Pennsylvania Emergency Management Agency
Pennsylvania National Guard
Pennsylvania Urban Search and Rescue Team
US Chemical Safety Board
US Coast Guard Atlantic Strike Team
US Postal Service, Postal Road Facility

Support Services:

Action Rental
Air Products & Chemicals, Inc.
Alvin H. Butz, Inc.
American Red Cross
Amey Clean Right
AMQUIP
Andy's Crane Services
Atlas Towing, Inc.
Deiter Bros. Fuel Company
E. PA Mountaineers Search and Rescue
First Union Bank, Lehigh Valley
Hechinger Home Project Center

LANTA
Lehigh County Communications Center
Lehigh Valley Int. Airport Com Center
MEDCOM
NE Search and Rescue
Northampton County Com Center
N. PA Goodwill
PP&L
Salvation Army
Stauffer Mfg. Company
Trexler-Haines Gas, Inc.

APPENDIX B

Maps of the Lehigh Valley Industrial Park

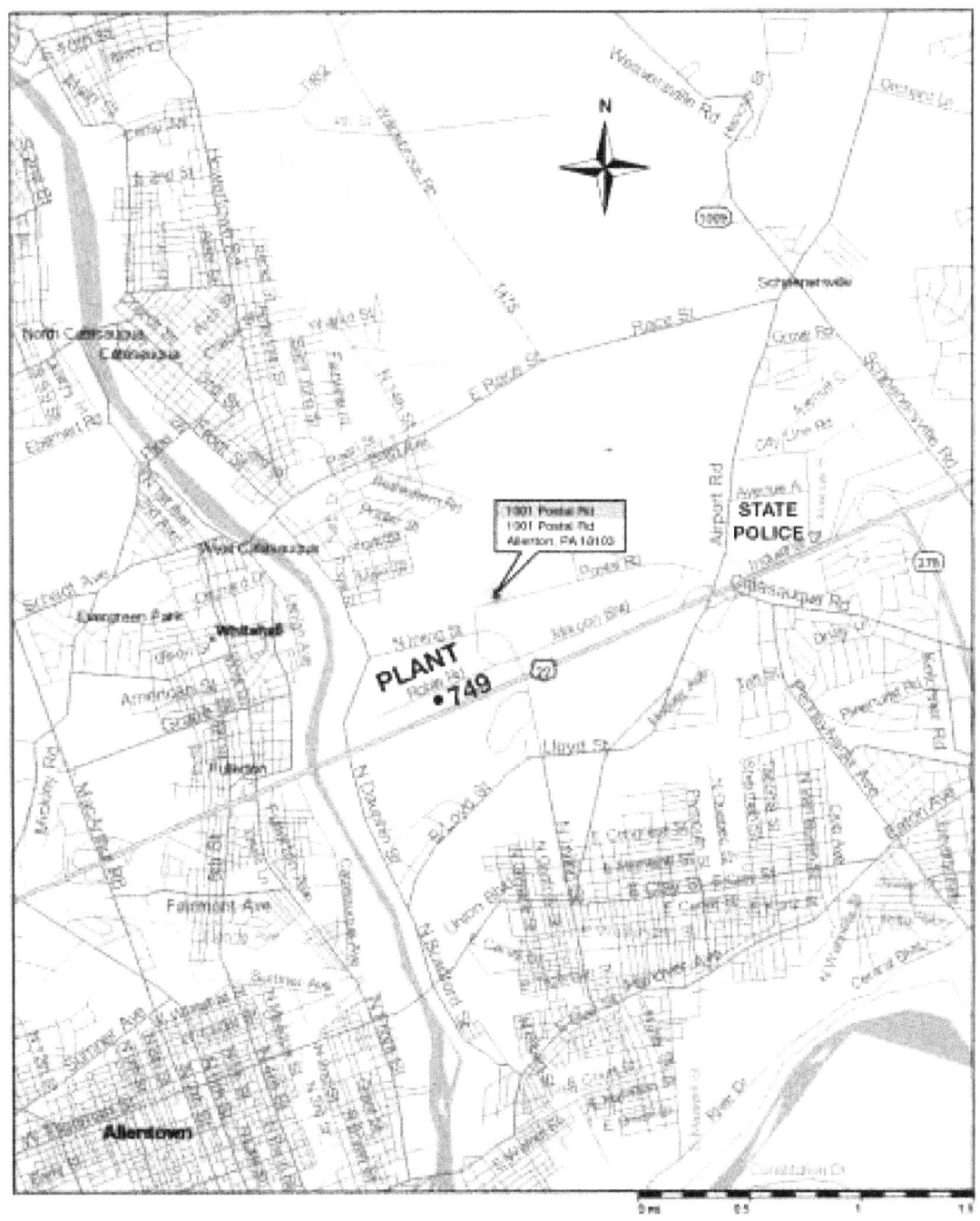

Lehigh Valley Industrial Park – Overall

Appendix B (continued)

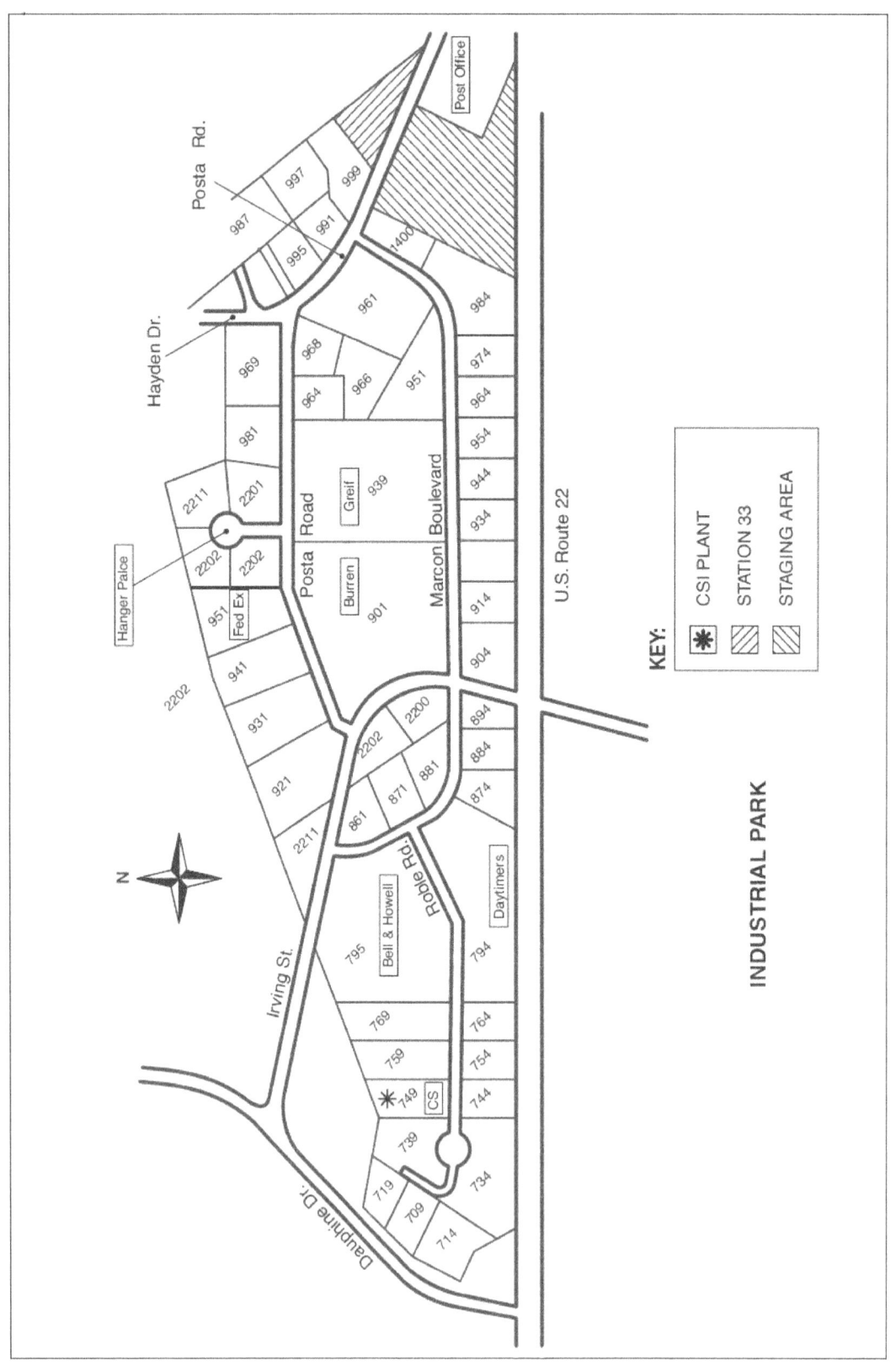

Lehigh Valley Industrial Park – Immediate Area

APPENDIX C

Hazardous Materials Guide for Hydoxylamine

HYDROXYLAMINE

Other Names: Oxammonium

WARNING! · EXPLOSIVE! CONTAINER MAY EXPLODE WHEN EXPOSED TO FIRE OR HEATED ABOVE 265°F! MAY EXPLODE AT LOWER TEMPERATURES IF MATERIAL IS EXPOSED TO AIR!

Hazards:	Description:
• May interfere with the body's ability to use oxygen • Very irritating to skin, eyes, nose and lungs; prolonged contact with skin can cause burns • Combustion or decomposition products upon heating include toxic nitrogen oxides	• White solid • No odor found • Melts at 90° F forming a colorless liquid • Initally sinks in water and is soluble in water • Nonflammable
Awareness and Operational Level Training Response: • Stay upwind and uphill • Determine the extent of the problem • BACK OFF! - Isolate a wide area around the area of release or fire and deny entry and call for expert help • For container exposed to fire evacuate the area in all directions because of the risk of explosion • Notify local health and fire officials and pollution control agencies • If material or contaminated runoff enters waterways, notify downstream users of potentially contaminated water	**Operational Level Training Response:** RELEASE, NO FIRE: • Cover material to protect from wind, rain or spray • Prevent runoff from entering sewers and waterways if it can be done safely well ahead of the release • Ventilate confined area if it can be done without placing personnel at risk FIRE: • Use extreme caution in approaching fire because material may explode without warning; no attempt should be made to fight fires except with unattended monitors using an agent appropriate for the burning material • Cool exposed containers with large quantities of water from unattended equipment or remove intact containers if it can be done safely • If cooling streams are ineffective (unvented container distorts, bulges or shows any other signs of expanding), withdraw immediately to a secure location

First Aid:
- Provide Basic Life Support/CPR as needed
- Decontaminate the victim as follows:
    - Inhalation - remove the victim to fresh air and give oxygen if available
    - Skin - remove and isolate contaminated clothing (including shoes) and wash skin with soap and large volumes of water for 15 minutes
    - Eye - rinse eyes with large volumes of water or saline for 15 minutes
    - Swallowed - do not make the victim vomit
- Seek medical attention
- For skin burns decontaminate with water and apply a clean dry dressing
- Note to physician: can cause methemoglobinemia; if symptoms indicate, methylene blue is the initial antidote

CAS: 7803-49-8

APPENDIX D

MSDS for Hydoxylamine

Concept Sciences, Inc.

50-HA™

Material Safety Data Sheet

| 1. | CHEMICAL PRODUCT AND COMPANY IDENTIFICATION |

PRODUCT NAME: 50-HA™

CHEMICAL NAME: HYDROXYLAMINE, 50 WT. % SOLUTION IN WATER

MANUFACTURER: Concept Sciences, Inc.
450 Allentown Drive
Allentown, PA 18103

FOR PRODUCT INFORMATION CALL:
Concept Sciences, Inc.
450 Allentown Drive
Allentown, PA 18103
610-435-4004

IN CASE OF EMERGENCY CALL:

CHEMTREC:	1-800-424-9300
Concept Sciences, Inc.:	1-610-435-4004
After Business Hours:	
(Dr. Chip Ward)	1-610-691-6889
(Brian Heath)	1-215-361-0198

| 2. | CHEMICAL PRODUCT IDENTIFICATION |

PRODUCT NAME: 50-HA™
OTHER NAME: HYDROXYLAMINE, 50 WT. % SOLUTION IN WATER
SYNONYM: OXAMMONIUM

COMPONENT	CAS NUMBER	WEIGHT %
Hydroxylamine	7803-49-8	48-52
Water	7732-18-5	48-52
Stabilizer	Trade Secret	<0.01

| 3. | HAZARDS IDENTIFICATION |

HAZARD PRECAUTIONARY STATEMENTS:
- Corrosive
- Corrosive to skin - causes burns
- Do not distill to dryness
- Excessive & extended heating may cause a "pressure build up" explosion within a closed container
- Decomposes in presence of heat and air. Decomposition is accelerated with transition metal ions (e.g.: Fe, Cu, Ni, et al)

Appendix D (continued)

Concept Sciences, Inc.

50-HA™

Material Safety Data Sheet

3.	HAZARDS IDENTIFICATION – (cont.)

POTENTIAL HEALTH HAZARDS:

SKIN: Corrosive to skin. Will cause burns to the skin. May cause allergic reactions.

EYES: Causes irritation and/or burns to the eyes

INHALATION: May cause irritation to the upper respiratory tract. Additional effects may be shortness of breadth, headache, dizziness, bluish skin color, convulsions and coma

INGESTION: Harmful if shallowed. May cause vomiting, digestive disorders, headache, dizziness and bluish skin color.

DELAYED EFFECT: May cause methemoglobinemia. May have potential mutagenic and teratogenic effects. Sensitizer.

Ingredient found on one of the OSHA designated carcinogen lists are listed below.

COMPONENT NAME	NTP STATUS	IARC STATUS	OSHA LIST
No ingredients listed in this section.			

4.	FIRST AID MEASURES

SKIN: In case of contact, immediately flush with copious amounts of soap and water for at least 15 minutes while removing contaminated clothing and shoes.

EYES: Assure adequate flushing of the eyes by separating the eyelids with fingers. Flush with water for at least 15 minutes. Seek medical attention.

INHALATION: Remove to fresh air. If not breathing, give artificial respiration. If breathing is difficult, give oxygen.

INGESTION: If swallowed, wash out mouth with water and swallow two to four glasses of water provided person is conscious. Induce vomiting. Call a physician.

GENERAL: Wash contaminated clothing before reuse. Discard contaminated shoes.

ADVICE TO PHYSICIAN: Test for methemoglobin. Treat according to symptoms.

Appendix D (continued)

Concept Sciences, Inc.

50-HA™

Material Safety Data Sheet

5.	FIRE FIGHTING MEASURES

FLAMMABLE PROPERTIES

FLASH POINT: . . . . None. (decomposes > 158° F (70°C)

FLASH POINT METHOD: . . . . Not applicable.

AUTOIGNITION TEMPERATURE: . . . . . Not applicable.

UPPER FLAME LIMIT (volume % in air): . . Not determined.

LOWER FLAME LIMIT (volume % in air): . . Not determined.

FLAME PROPAGATION RATE (solids): . . . Not applicable.

OSHA FLAMMABILITY CLASS: . . . . . . Not applicable.

EXTINGUISHING MEDIA:

- Flood with water.
- Use halon, carbon dioxide, dry chemical powder or appropriate foam or earth if water is unavailable.

SPECIAL FIREFIGHTING PROCEDURES:

- Wear self-contained breathing apparatus and full protective clothing to prevent contact with skin and eyes.
- Fight fire from safe distance as material MAY be explosive if water has evaporated.

UNUSUAL FIRE AND EXPLOSIONS HAZARDS:*

- Emits toxic fumes under fire conditions (e.g. N_xO_y, NH_3)
- Container explosion may occur under fire conditions.
- Danger of fire hazard when exposed to extended and excessive heat, or flame or oxidizers.
- Combustible rags/cloths used to soak up small spills may smolder considerably when dried in air or under heat or direct sunlight. Such rags must be thoroughly rinsed several times with water, squeezed and rinsed again before discarding.

* 50-HA™ as a water solution presents no danger of fire or explosion by itself. 50-HA™ can be safely used and handled as a water solution. Danger of a "pressure build-up" explosion exists as water is removed or evaporated and HA concentration approaches levels in excess of about 70% within a closed container.

Appendix D (continued)

Concept Sciences, Inc.

50-HA™

Material Safety Data Sheet

6.   ACCIDENTAL RELEASE MEASURES

- Evacuate area
- Wear self-containing breathing apparatus, rubber boots and heavy rubber gloves
- Cover with solid citric acid and inert absorbent, pick up and place in a closed container and hold for waste disposal.
- 50-HA™ can be neutralized (slowly because of exotherm) with approximately a 20% aqueous hydrochloric acid solution to a pH of approximately 7.
- Ventilate area and wash spill site wit water after material pick-up is complete. Refer to Section 5 for disposal of rags/cloths used in material pickup.

7.   HANDLING AND STORAGE

Normal Handling:
- (Always wear recommended personal protective equipment.)
- Ensure adequate local exhaust ventilation.
- Avoid contact with skin, eyes and clothing.
- Wash thoroughly after handling.
- Do not eat or drink while handling.

Storage Recommendations:
- Keep containers tightly closed
- Protect from physical damage and atmospheric moisture.
- Store in cool, noncombustible buildings away from oxidizing materials and incompatible substances.
- Avoid conditions where material will be exposed to excessive heat (>70°C) and exposure to transition materials (e.g., iron, copper, nickel, aluminum, et. al.)

8.   EXPOSURE CONTROLS/PERSONAL PROTECTION

Engineering Controls:
Ensure adequate local exhaust ventilation. Ventilation equipment should be explosion proof.

Personal Protective Equipment:
SKIN PROTECTION
- Rubber Gloves
- Apron or long sleeve coveralls
- Boots
- Arm protection

Appendix D (continued)

Concept Sciences, Inc.

50-HA™

Material Safety Data Sheet

8.   EXPOSURE CONTROLS/PERSONAL PROTECTION (cont'd)

EYE PROTECTION
Chemical splash goggles

Respiratory Protection:
Fullface respirator with mist/amine type cartridge if exposed to vapors

Additional Recommendations:
- Eye wash station
- Wash thoroughly after handling
- Avoid prolonged or repeated exposure
- Keep 50-HA™ in tightly closed containers with vented caps
- In laboratory, use only in exhausted chemical fume hood

EXPOSURE GUIDELINES:

COMPONENT NAME	ACGIH TLV	OSHA PEL	OTHER LIMIT
No component listed in this section.			

9.   PHYSICAL AND CHEMICAL PROPERTIES

Appearance:	Clear, colorless liquid
Molecular Weight:	33.04 (100% hydroxylamine)
Chemical Formula:	NH_2OH/H_2O
Odor:	Sl. Ammoniacal
Specific Gravity (water=1.0)	1.11 to 1.13
Solubility in Water (weight %)	Soluble in all proportions
ph: (50% w/w water solution)	-10.5
Boiling Point:	105-107°C
Melting Point:	N/A
Vapor Pressure:	10 mm Hg @ 47°C (for 100% hydroxylamine)
Vapor Density (air=1.0)	>1.0 (no quantitative value available)
% Volatiles:	≥99.0
Flash Point	None

(Flash point method and additional flammability data are found in Section 5.)

Concept Sciences, Inc.   MSDS – 50-HA™   Page 5 of 11
Issue Date: 3/19/98 Rev. 5   Print Date: 09/02/98

Appendix D (continued)

Concept Sciences, Inc.

50-HA™

Material Safety Data Sheet

---

10. STABILITY AND REACTIVITY

Stability:
- Stable

Conditions to avoid:
- Do not distill to dryness
- Do not heat to dryness
- Avoid strong acids or bases
- Avoid heavy metal or transition metal contamination, (e.g.; Cu, Fe, Ni in particular)

Incompatibilities:
- Heat (excessive or prolonged which will remove solution in water)
- Oxidizing agents
- Potassium Dichromate
- Chromium Trioxide
- Zinc
- Calcium
- Copper or Copper Compounds
- Iron or Iron Compounds
- Ammonia
- Phosphorous Halides
- Carbonyls
- Pyridine
- Hypochlorites
- Transition Metals, ions, compounds

Hazardous Combustion or Decomposition Products
- Toxic fumes of: Nitrogen oxides
                           Ammonia

Hazardous Polymerization
- Will not occur

---

11. TOXICOLOGICAL INFORMATION

ACUTE EFFECTS: *Harmful if swallowed or absorbed through skin*

Routes of Entry/Effects:
- **INGESTION:** Absorption into the body through injection into mice has lad to the formation of methemoglobin which in sufficient concentration can cause cyanosis. Onset may be delayed 2 to 4 hours or longer.

Appendix D (continued)

Concept Sciences, Inc.

50-HA™

Material Safety Data Sheet

11.   TOXICOLOGICAL INFORMATION (cont'd)

- **INHALATION:** May cause spasm, inflammation and edema of the larynx and bronchi, chemical pneumonia and pulmonary edema.

- **SKIN:**   Dermal Irritation/Corrosion:
    - Corrosive to skin when applied for a 4 hour period.
    - Irritating to skin when applied for a 1 hour period.
    - Non-irritating to skin when applied for a 3-minute period.

Stability:
- Stable
- Coughing
- Wheezing
- Hot or Sore Throat
- Laryngitis
- Shortness of breath
- Headache
- Nausea
- Vomiting
- Slight to moderate skin irritation, depending on degree of exposure.

Target Organs: Blood, Central Nervous System

Conditions to avoid:
- NTP   Not listed
- IARC Monographs: Not listed
- OSHA Regulated: No

Hydroxylamine has been reported to be mutagenic in vitro to lower organisms.

Medical Conditions Generally Aggravated by Exposure:
- Repeated exposure may cause dermatitis

Delayed (Subchronic and Chronic) Effects:
- Hydroxylamine has been reported to produce developmental effects in rabbits, but only when embryos were directly exposed to the chemical.
- No developmental effects were found in studies with rats and calves.
- Hydroxylamine is mutagenic in vitro, but not in vivo.
- Hydroxylamine was not carcinogenic in several studies.

Appendix D (continued)

Concept Sciences, Inc.

50-HA ™

Material Safety Data Sheet

---

11.   TOXICOLOGICAL INFORMATION (cont'd)

RTECS #: NC2975000

Toxicity Data:

IPR-RAT	LD50:59	MG.KG	CNREA8	26,1448,66
SCU-RAT	LD50:29	MG/KG	JPETAB	119,444,57
IPR-MUS	LD50:60	MG/KG	JPETAB	165,30,69
UNR-MUS	LD50:175	MG/KG	NJCAAI	6,160,52
ORAL-RAT	LD50:190	MG/KG		
ORAL-RAT-FEMALE	LD50:120	MG/KG		
ORAL-RAT-MALE	LD50:248	MG/KG		

Only selected registry of toxic effects of chemical substances (RTECS) data is presented here. See actual entry in RTECS for complete information.

---

12.   ECOLOGICAL INFORMATION

Hydroxylamine is harmful to aquatic life in very low concentration. Do not allow 50-HA™ direct entry into sewers or waterways.

---

13.   DISPOSAL CONSIDERATIONS

RCRA

Unused 50-HA™ is NOT considered an RCRA hazardous waste.

Other Disposal Considerations:

Although 50-HA™ is not a hazardous waster (pH (~10.5), it is not suitable for disposal down the drain. See Section 6 for neutralization procedures. Waste 50-HA™ should be treated with hydrogen peroxide before release to sewer. Check with local authoritities before any such disposal method is employed.

The information offered here is for the product as shipped. Use and/or alterations to the product such as mixing with other materials may significantly charge the characteristics of the material and alter the RCRA classification and the proper disposal method.

50-HA™ may be dissolved or mixed with a combustible solvent and burned in a chemical incinerator equipped with an after burner and scrubber.

Observe all federal, state and local environment regulation.

Appendix D (continued)

Concept Sciences, Inc.

50-HA™

Material Safety Data Sheet

14.   TRANSITION INFORMATION

DOT Description: Corrosive Liquid, Toxic N.O.S. (Hydroxylamine),
8, UN2922, PGIII

NOS Description: Hydroxylamine

15.   REGULATORY INFORMATION

RCRA
TSCA Inventory Status: Components are listed on the TSCS Inventory
Other TSCA Issues: None

Sara Title III/CERCLA
"Reportable Qualities" (RQ's) and/or "Threshold Planning Quantities" (TPQ's) exist for the following components:

Component Name   SARA/CERCLA RQ[40 CFR 302.4(a)]   SARA EHS TPQ (lb)
No components listed in this section.

Sara 302/313 Components:
None listed.

Spills or releases resulting in the loss of any ingredient at or above its RQ requires immediate notification to the National Response Center [(800) 424-8802] and to your Local Emergency Planning committee.

Section 311/312 Hazard Class: (40 CFR370.2)
Immediate (X)   Delayed ( )   Fire* ( )   Reactive* ( )   Sudden Release ( )
of Pressure

* (50-HA) does not meet Fire or Reactive criteria defined in 29 CFR, Sec. 1910.1200. 50-HA may decompose in time with release of flame retardant gases if contaminated by certain transition metals, particularly Fe, Ni and Cu. Elevated temperature may hasten decomposition rate of contaminated material

State Right-To-Know
In addition to the components found in Section 2, the following are listed for state right-to-know purposes
NONE

Concept Sciences, Inc.   MSDS – 50-HA™   Page 9 of 11
Issue Date: 3/19/98 Rev. 5   Print Date: 09/02/98

Appendix D (continued)

Concept Sciences, Inc.

50-HA™

Material Safety Data Sheet

15.	REGULATORY INFORMATION

Additional Regulatory Information:

OSHA Process Safety Management Standard (20 CFR 1910.119): (50-50™) as a 50% aqueous solution does not meet hazardous or highly hazardous chemical criteria as defined in 29 CFR 1910.119 as 50% aqueous solution.

Also, (50-Ha™) as a 50% aqueous hydroxylamine solution does not meet the criteria for a "hazardous substance" as defined in 49 CFR 171.8 of the Hazardous Materials Guide." It is not listed in Appendix A to 172.101, Table Hazardous Substances other than Radionuclides

WHMIS Classification (Canada):

Not determined

Foreign Inventory Status:

Not determined

Reviews, Standards and Regulations:

- OEL=MAK
- NOHS 1974: HZD 84783; NIS 3; TNF 116; NOS 5; TNE 1722
- NOES 1983: HZD 84783; NIS 5; TNF 112; NOS 9; TNE 12689; TFE 1481
- EPA GENETOX PROGRAM 1988, POSITIVE: CELL TRANSFORM.–RLV F344 RAT EMBRYO
- EPA GENETOX PROGRAM 1988, POSITIVE: N CRASSA-FORWARD MUTATION; S POMBE-FORWARD MUTATION
- EPA GENETOX PROGRAM 1988, NEGATIVE: IN VITRO CYTOGENETICS-HUMAN LYMPHOCYTE
- EPA GENETOX PROGRAM 1988, NEGATIVE: D MELANGASTER-WHOLE SEX CHROM. LOSS
- EPA GENETOX PROGRAM 1988, INCONCLUSIVE: CARCINOGENICITY-MOUSE-RAT
- EPA GENETOX PROGRAM 1988,INCONCLUSIVE: D MELANOGASTER-NONDISJUNCTION
- EPA GENETOX PROGRAM 1988,INCONCLUSIVE: D MELANOGASTER SEX-LINKED LETHAL
- EPA TSCA TEST SUBMISSION (TSCATS) DATA BASE, OCTOBER 1996

Appendix D (continued)

Concept Sciences, Inc.

50-HA ™

Material Safety Data Sheet

16.   OTHER INFORMATION

Other Information:
- **NFPA Rating**
  - Health – 2
  - Flammability – 0
  - Reactivity – 3

The information above is believed to be accurate and represents information given in good faith. However, we make no warranty of merchantability or any other warranty, express or implied, with respect to such information, and we assume no liability resulting from its use. Users should make their own investigations to determine the suitability of the information for their particular purposes. In no way shall concept Sciences, Inc. be liable for any claims, losses, or damages of any third party or for lost profits or any special, indirect, incidental, consequential or exemplary damages, howsoever arising, even if Concept Sciences, Inc. has been advised of the possibility of such damages.

```
DISTRIBUTION:    CHEMTREC
                 Allentown Fire Department
                 MSDS BOOK     -QA LAB
                 MSDS BOOK     -PROD. DEPT.
                 MASTER FILE   -L. Frank
```

Issue Date: 3/19/00	Rev. # 5	Approval: C. Ward (Original Signature on File) 3/19/98

Site Plans

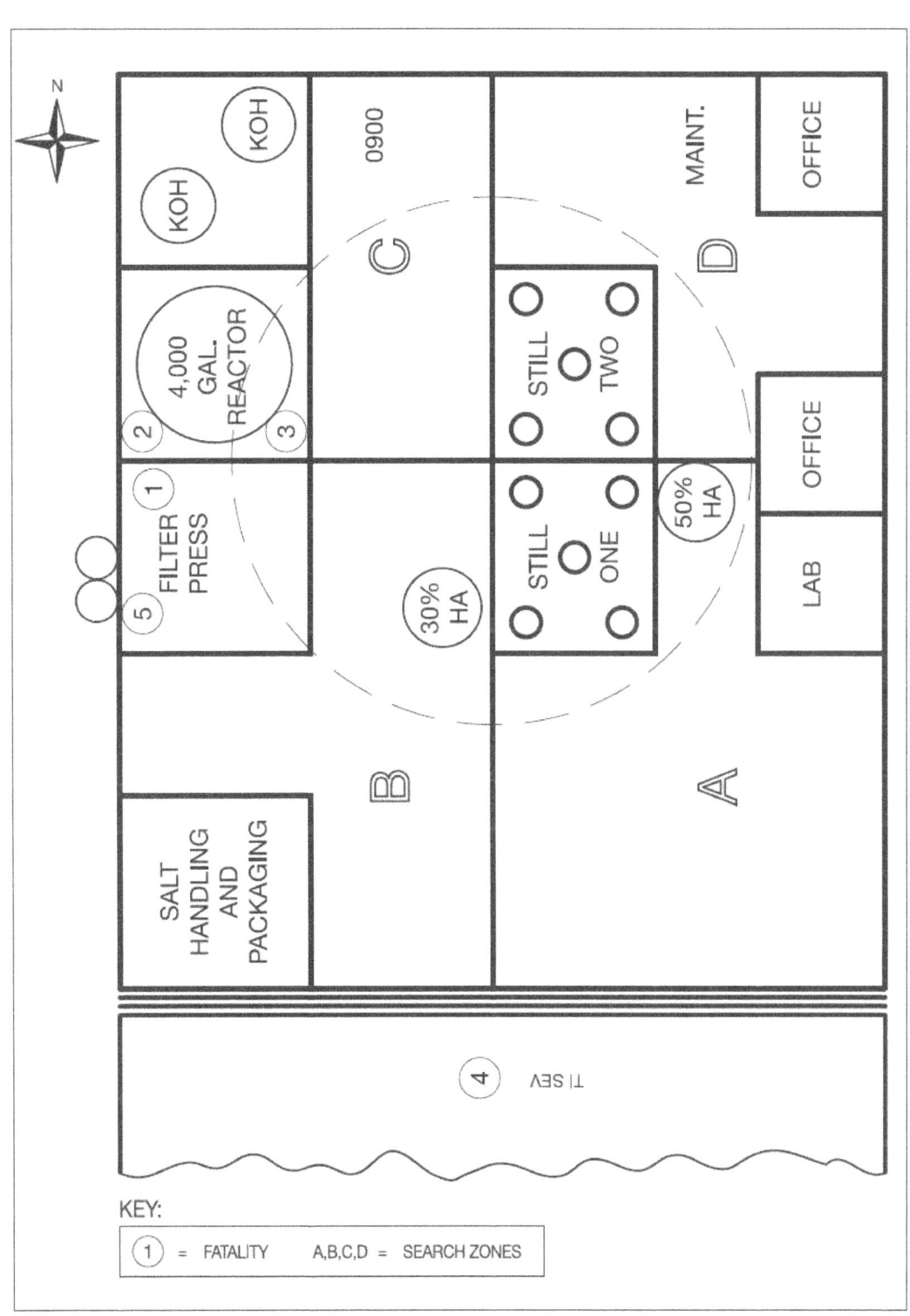

CSI Site Plan

KEY:

① = FATALITY   A,B,C,D = SEARCH ZONES

Appendix E (continued)

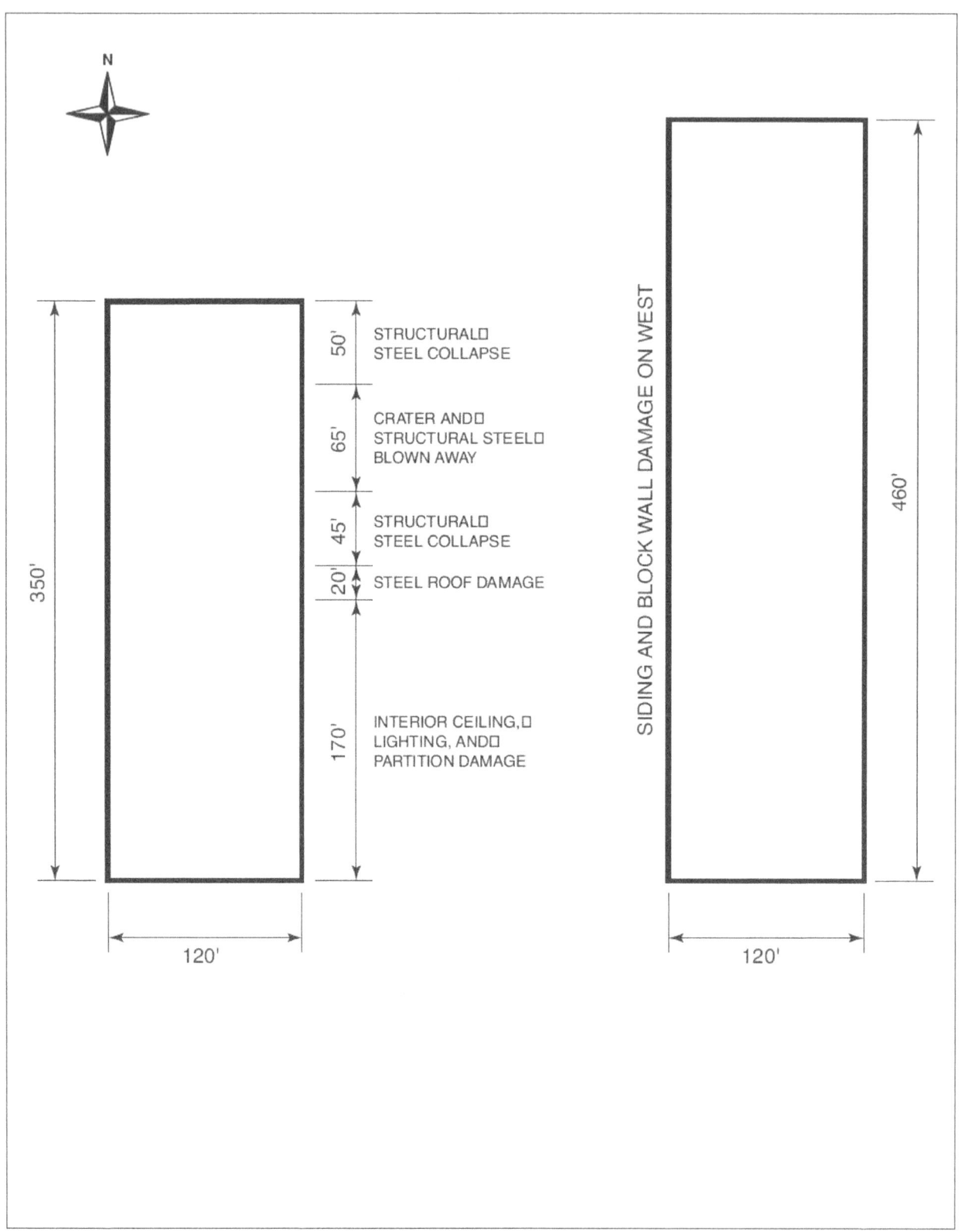

749 and 759 Roble Road

APPENDIX F

Photographs

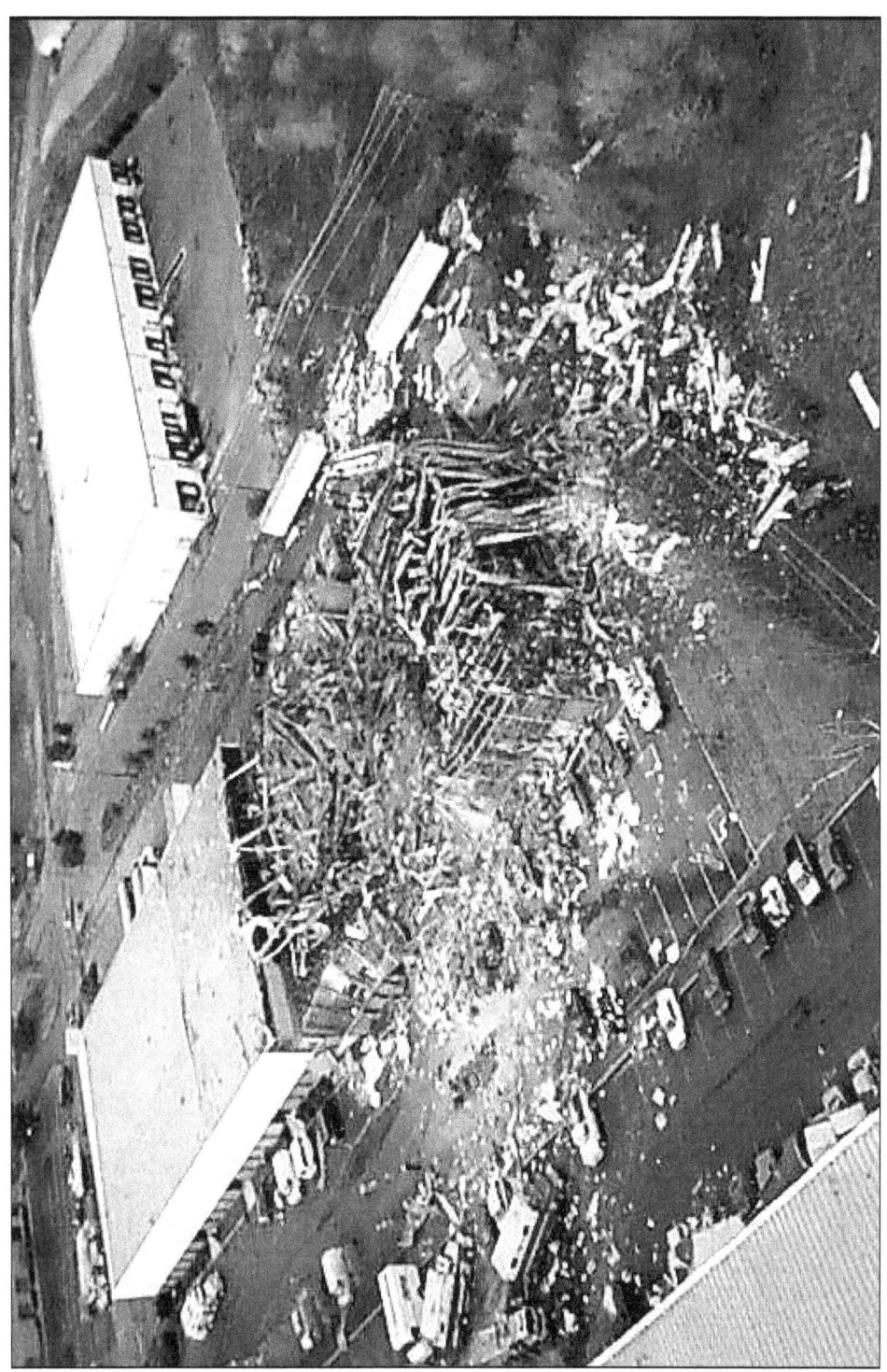

Source: David Lesak

1. Aerial view of the CSI Plant after the explosion.

Appendix F (continued)

Source: David Lesak

2. View of the extent of the destruction.

Appendix F (continued)

Source: David Lesak

3. View of the crater created by the explosion in the floor of the CSI plant.

Appendix F (continued)

Source: Steve Martin

4   View of the destruction to the side of the CSI building.

Appendix F (continued)

Source: Steve Martin

5. CSI Command Post and Operations Sector (Note date should read 2/20/99)

APPENDIX G

Mechanics of an Explosion

Definitions

Brisance: The sharp, shattering effect upon the surroundings, determined by velocity of the detonation wave.

Deflagration: Strictly speaking an exothermic reaction which propagates from the burning gases to the unreacted material by conduction, connection, or radiation, also a rapid burning. Deflagrations are associated with low order explosives. The rate is subsonic.

Detonation: To explode with sudden violence, associated with high order explosives. An exothermic (giving off heat) reaction that is characterized by the presence of a shock wave in the material that establishes and maintains the reaction. Rate is supersonic. Principal heating mechanism is one of shock compression; temperature rise is directly associated with the intensity of the shock wave.

Explosion: A bursting or violent expansion as a result of a sudden production or release of pressure. It is always accompanied by a loud noise, high temperature, and usually by a large volume of gas. There are three types of explosions: Chemical, Mechanical, and Nuclear.

Initiator: A mechanism or action that induces or starts an explosion or the chain of events leading to an explosion.

Shrapnel: Fragments from an explosion.

Source: R.R. Lenz (1965). *Explosives and Bomb Disposal Guide.* Springfield, IL: Charles C. Thomas Publisher.

The different ways that something may explode.

Chemical explosions:

- **Rapid oxidation:** the most common type of explosion. Oxidation is a chemical reaction that takes place with oxygen. If a few ounces of gasoline are ignited in a container the vapor will burn at a normal rate. If the same amount of gasoline is completely vaporized in air it can burn in a fraction of a second.

- **Runaway polymerization:** Polymerization is the chemical combination of smaller molecules into much larger ones. If this process is uncontrolled,, unstable molecules form chains and release energy in doing so. The heat produced from this reaction can cause an explosion.

- **Decomposition of molecules:** molecules breaking apart into simpler fragments can also release energy that can cause an explosion.

Mechanical explosions:

- **Pressure relief:** when a pressurized vessel is exposed to fire or whenever steam is trapped in a vessel or container the vessel may rupture in order to relieve the excess pressure.

Nuclear explosions:

- **Nuclear fision:** when the nuclei of certain atoms split.

- **Nuclear fusion:** after a complex chain of events multiple atoms fuse into a single atom.

<u>Source</u>: J. H. Meidl (1970). Explosive and Toxic Hazardous Materials. Beverly Hills, CA: Glencoe Press.